T0132782

FLORA OF TROPICAL EAST AFRICA

ALANGIACEAE

B. Verdcourt

(East African Herbarium)

Trees or shrubs, often spiny. Leaves alternate, exstipulate, petiolate, entire or lobed, often asymmetric at the base. Flowers hermaphrodite, regular, in few to many-flowered axillary cymes. Calyx truncate or divided into 4–10 lobes, or denticulate. Petals 4–10, linear, valvate, often joined at the extreme base. Stamens equal in number to the petals and alternating with them or up to four times as many, free or slightly joined at the base. Ovary inferior, 1–2-celled, ovules solitary and pendulous in the cells. Style simple, clavate or 2–3-lobed. Fruit a drupe crowned with sepals and disc, 1–2-seeded ; seeds albuminous.

The family has only the following genus.

ALANGIUM

Lam., Encycl. 1 : 174 (1783)

Characters of the family.

Species 1 belongs to the section *Angolum* Baill. and species 2 to the section *Marlea* Baill.

Stamens over 15, more than the petals, flowers 18 mm.
long and 4 mm. wide in more or less sessile inflorescences . . . . 1. *A. salviifolium*
subsp. *salviifolium*

Stamens 5–8 the same number as the petals, flowers
8–16 mm. long and 1·5–3·5 mm. wide in pedunculate inflorescences . . . . . . 2. *A. chinense*

1. **A. salviifolium** (*L. f.*) *Wangerin* in E.P. IV. 220b : 9 (1910). Type : Ceylon, *Koenig* (LINN–SM)

Medium-sized tree or climbing shrub with rough light brown bark. Branchlets grey- or purple-brown, often with strong spines up to 1·2 cm. long, pubescent or glabrous. Leaves variable, cuneate or rounded at the base, rounded or acute at the apex, at first pubescent later glabrous, 3–23 cm. long and 1·4–9 cm. wide, petiole up to 1·5 cm. long ; lateral nerves 3–9, venation openly reticulate, prominent below. Flowers cream with a slight orange tinge, 5–10-merous, 13–31 mm. long, 1–17 in almost sessile inflores-cences, pedicels 2–8 mm. long. The pedicels and cylindrical buds are velvety with golden-brown hairs. Calyx tube urceolate 0·75–2·5 mm. long, lobes triangular 0·25–1·5 mm. long. Petals ligulate, densely pubescent outside, internally pubescent or glabrous, 12–28·5 mm. long and 1–2·3 mm. wide. Stamens 10–32, anthers narrow, 5–14 mm. long. Style glabrous,

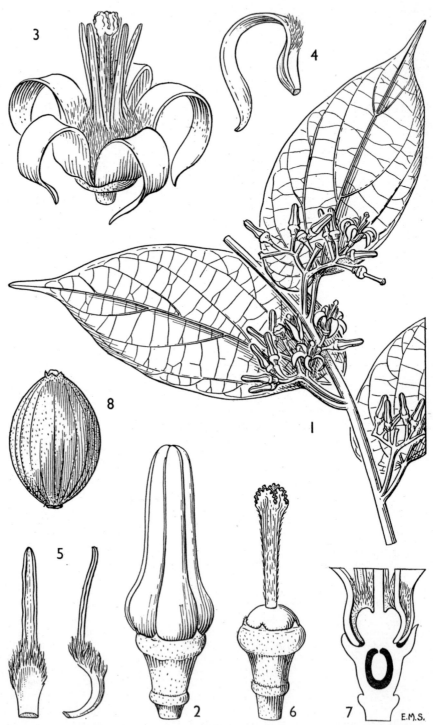

FIG. 1. *ALANGIUM CHINENSE*—**1**, flowering branch, from below, × 1 ; **2**, flower-bud, × 6 ; **3**, flower, ×6 ; **4**, petal, × 6 ; **5**, stamens, × 6 ; **6**, flower with petals and stamens removed, showing pistil, × 6 ; **7**, part of flower in longitudinal section (diagrammatic) to show placentation and insertion of petals and stamens, × 6 ; **8**, fruit, × 4. 1, from *Greenway* 3309, 2–7, from *Dale* (in spirit) ; 8, from *Ujor* 30389A.

slightly dilated above, stigma capitate or conical, somewhat lobed, 8·5–27·5 mm. long. Fruit ellipsoidal, shortly pubescent, 9–22·5 mm. long, one-celled.

subsp. salviifolium

Medium-sized tree up to 18 m. tall with trunk 0·6 m. in diameter, branchlets at first pubescent, later glabrous, often with strong spines up to 1·2 cm. long. Leaves elliptic to oblong-obovate, cuneate at the base, rounded or acute at the apex, 3–9·5 cm. long and 1·7–4·3 cm. wide ; lateral nerves 3–6. Flowers usually 7-merous in African plants, 18 mm. long, 2–5 in almost sessile inflorescences. Calyx tube 2 mm. long, lobes 1 mm. long. Petals 16–17·5 mm. long. Stamens usually 17 in African plants, anthers 9–10 mm. long. Style and stigma together 14 mm. long.

KENYA. Teita District : Taveta Forest, Nov. 1937 (fl.), *Dale* 3763 !
TANGANYIKA. Kilosa District : Vigude Forest, Oct. 1952 (fl.), *Semsei* 977 ! and Dec. 1952 (fr.), *Semsei* 1060 !
DISTR. **K**7 ; **T**6 ; Comoro Islands, India to Indo-China and China
HAB. Lowland rain-forest, 500–750 m.

SYN. *Grewia salviifolia* L. f., Suppl. : 409 (1781)
    A. salviifolium (L. f.) Wangerin subsp. *decapetalum* (Lam.) Wangerin in E.P. IV. 220b : 11 (1910) ; Bloembergen in Bull. Jard. Bot. Buitenzorg (series 3) 16 : 152 (1939)
    A. salviifolium (L. f.) Wangerin subsp. *hexapetalum* (Lam.) Wangerin in E.P. IV. 220b : 9 (1910), quoad nom., excl. descr.

NOTE. Wangerin recognized two subspecies, *hexapetalum* (Lam.) Wangerin and *decapetalum* (Lam.) Wangerin. The Comoro forms he attributed to the former. Bloembergen saw Lamarck's types and showed that both were the same taxon— *decapetalum* and that the form which Wangerin called *hexapetalum* should be known as subsp. *sundanum* (Miq.) Bloem.; this subsp. does not occur within the area covered by this Flora. He also disagrees with Wangerin and refers the Comoro material to subsp. *decapetalum* but admits that it is rather isolated. Linnaeus' type is subsp. *decapetalum* which taxon must therefore be called subsp. *salviifolium*. The present records from tropical Africa are apparently the first for the continent. Further material may show it to be varietally distinct.

2. **A. chinense** (*Lour.*) *Harms* in Ber. D. Bot. Ges. 15 : 24 (1897); Bloembergen in Bull. Jard. Bot. Buitenzorg, ser. 3, 16 : 169 (1939). Type : Cochinchina, *Loureiro* (P, holo., BM, iso.)

Moderate-sized to large tree, 9–24 m. tall, bark smooth and grey. Branchlets purple-brown, never spiny, at first pubescent later glabrous. Leaves very variable, narrowly elliptic to broadly ovate, usually ovate, cuneate, truncate or subcordate at the base, asymmetric, often markedly so, acuminate at the apex, entire when adult but seedling leaves and leaves from coppice shoots are often palmately lobed and much larger than adult leaves. Lamina 4–19 cm. long and 2·5–10 cm. wide (up to 25 × 27 cm. in juvenile leaves), at first pubescent, later glabrescent ; petiole 0·5–2·5 cm. long, velvety pubescent or glabrous ; main lateral nerves 4–6, the basal ones confocal and sharply ascending so that the lamina is 5–7-nerved from the base, tertiary venation close, approximately parallel and at right angles to the costa. Flowers golden pubescent, white, cream or yellow, sweetly scented, 8–16 mm. long and 1·5–3·5 mm. wide, 5–8-merous, 3–23 in pedunculate 1–4-branched pubescent or glabrescent inflorescences, peduncles 0·4–1·8 cm. long, secondary branches 0·1–2·6 cm. long and pedicels 3–5·5 mm. long. Buds cylindrical, usually swollen at the base. Calyx-tube cylindrical or funnel-shaped, 1·5 mm. long, limb spreading, 1 mm. tall, denticulate. Petals ligulate, 8–13·5 mm. long and 1–1·7 mm. wide, pubescent outside, glabrescent inside save for a densely hirsute area near the base. Stamens the same number as the petals, anthers narrow, 5–9 mm. long. Style pilose, cylindrical, clavate apically, stigma lobed, together 7–10 mm. long. Fruit laterally compressed, globose or ellipsoidal, rather costate in the dry state, basally rounded, apically rather tapering, minutely pubescent, 1–2-celled, 8–10 mm. long, 6·5–9 mm. wide (maximum) and 4–5·5 mm. wide (minimum). Fig. 1.

UGANDA. Kigezi District : Ishasha Gorge, May 1950, *Purseglove* 3415 ! ; Mbale District : Elgon, Apr. 1927, *Snowden* 1079 !
KENYA. North Kavirondo District : Kakamega Forest, June 1934, *Dale* 3256 ! ; Meru, 1922, *Fries* 1811 !
TANGANYIKA. Moshi District : Kilimanjaro, Mashati, Jan. 1929, *Haarer* 1744 !; Tanga District : E. Usambara Mts., Mlessa, Dec. 1932, *Greenway* 3309 ! ; Rungwe District : Rungwe Mt., Mwangoka, *Stolz* 1057 !
DISTR. U2, 3 ; K4, 5 ; T2, 3, 6, 7 ; Belgian Congo, Northern Rhodesia, Angola and Portuguese Congo, Cameroons, Fernando Po. Also from India to China, Indo-China, Japan, Java, Philippine and Sunda Islands
HAB. A pioneer species in partly cleared areas of lowland and upland rain-forest, 750–2000 m.

SYN. *Stylidium chinense* Lour., Fl. Cochinchin. 1 : 221 (1790)
    Marlea begoniifolia Roxb., Pl. Corom. 3 : 80, t. 283 (1820). Type : Roxburgh's drawing (No. 2228) of a cultivated plant probably from Assam, Khasia Hills, Silhet (K, holo.)
    A. begoniifolium (Roxb.) Baill., Hist. Pl. 6 : 270 (1876) subsp. *eubegoniifolium* Wangerin in E.P. IV. 220b : 21 (1910) ; F.W.T.A. 1 : 519

NOTE. The combination *A. chinense* (Lour.) Rehder in Pl. Wilson., 2 : 552 (1916), frequently used, e.g. T.T.C.L.: 22 (1949) and I.T.U., ed. 2 : 4 (1952), is long antedated by Harms' combination. The very extensive extra-African synonymy is given by Bloembergen loc. cit. This species is remarkably variable in the shape of the leaves, particularly at the base, the relative lengths of the parts of the inflorescence and the indumentum. Bloembergen discusses the variation of the species throughout its range.

Doubtful Species

A. ? **kenyense** *Chiov.*, Racc. Bot. Miss. Consol. Kenya : 47 (1935).   Type : Kenya, Meru, *Balbo* 674 (TOM, holo. !)

This poorly prepared specimen is sterile and consequently its determination is not possible. It is probably a juvenile state of *A. chinense* (Lour), Harms, a species known to occur in this locality.

INDEX TO ALANGIACEAE